"Ryan doesn't look at ... ...lo, and this book is no different. Unusual perspectives are what make the world more interesting."

Douglas Coupland
Author of *Generation X*
Canadian novelist and artist

"Ryan's insights on social media's impact on business and society, coupled with his practical guide for navigating commerce in the digital age, make this book a must-read for any business leader."

Michele Romanow
"Dragon," CBC *Dragons' Den*
Cofounder and president, Clearbanc

"[*Saving Social* provides] an honest look at how social media has shaped business and human behavior from one of the industry's most-respected pioneers, with valuable insights into where the future of commerce is heading."

Lewis Howes
New York Times best-selling author
Host, *The School of Greatness*

"[*Saving Social* is] a holistic guide to navigating social media and its impact on business and society from one of the most prolific thought leaders in the industry."

Dharmesh Shah
Cofounder and CTO, HubSpot

"[*Saving Social* is] an insightful guide for understanding social media's long-term impact on business and preparing for the ever-changing landscape. Every business leader should read this."

Jeremiah Owyang
Tech analyst, Kaleido Insights

SAVING SOCIAL

RYAN HOLMES

SAVING SOCIAL

The Dysfunctional Past and
Promising Future of Social Media

Published by Advantage, Charleston, South Carolina.
Member of Advantage Media Group.

ADVANTAGE is a registered trademark, and the Advantage colophon is a trademark of Advantage Media Group, Inc.

Printed in the United States of America.

10 9 8 7 6 5 4 3 2 1

ISBN: 978-1-64225-166-1
LCCN: 2020919134

Cover design by David Taylor. Layout design by Carly Blake.

This publication is designed to provide accurate and authoritative information in regard to the subject matter covered. It is sold with the understanding that the publisher is not engaged in rendering legal, accounting, or other professional services. If legal advice or other expert assistance is required, the services of a competent professional person should be sought.

Advantage Media Group is proud to be a part of the Tree Neutral® program. Tree Neutral offsets the number of trees consumed in the production and printing of this book by taking proactive steps such as planting trees in direct proportion to the number of trees used to print books. To learn more about Tree Neutral, please visit www.treeneutral.com.

Advantage Media Group is a publisher of business, self-improvement, and professional development books and online learning. We help entrepreneurs, business leaders, and professionals share their Stories, Passion, and Knowledge to help others Learn & Grow. Do you have a manuscript or book idea that you would like us to consider for publishing? Please visit **advantagefamily.com** or call **1.866.775.1696**.

This book is dedicated to:

The Hootsuite customers who helped so much in the inspiration for this book.

The amazing Hootsuite team of whom I had the privilege of leading for so many years.

Remy, Jennifer, and Sabrina who were instrumental in the concepting, creating, editing, and promoting of this book ... basically all of it, thank you.

My Bianca and Paola, who are my everything.

Onwards, Ryan.

CONTENTS

Falling Apart, Coming Together

As I write this from my corner of our newly restricted world, society, business, and culture at large are in a process of mass recalibration. Companies built on the currency of physical connection, from retail shops to business offices, are struggling to find their digital footprints. Consumers, cautiously emerging from self-isolation, are adjusting to staggered lines and sanitized browsing in socially distanced shops. Parents, after months of managing homeschooling amid job loss or work-from-home constraints, are forgoing math lessons for discussions with their children on racial injustice. Monuments have come down, as have the legacy systems of the forefathers they depicted. Nothing is as it was. Nor should it be.

When I set out to write this book in the prepandemic context of 2019, I had a theory. Having spent more than a decade riding the culturally shifting wave of social media, I'd witnessed a definitive cycle. I'd experienced the heady beginnings of social media—a time when people marveled at the realization they could reconnect with old high school crushes, directly reach celebrities and respected industry leaders, and broadcast personal opinions globally.

I built a business helping companies capitalize on the paradigm shift social media brought into the corporate world. Suddenly, businesses everywhere had access to laser-targeted marketing, unfiltered feedback from real customers, convenience, transparency, and the ability to authentically connect with stakeholders while scaling operations at extraordinary speed. It was a game changer.

As the medium became more integrated into both business and mainstream culture, I watched euphoric novelty give way to growing concerns, and even hysteria, as worries over data collection and privacy began to mount. Fake news and trolls started to infiltrate the once collegial space of social media, and countless news cycles were dedicated to reporting on the ill effects of Cambridge Analytica and a sea of opposing views on how the government, if at all, should intervene.

Fast-forward to today, and everything from the ethics behind algorithms to the legal shield social networks use to protect themselves from liability is being discussed at the highest level of government. To many, it may appear as though the industry is coming undone, but what if, in fact, it's just starting to come together? You need only look to past storytelling technologies—radio, TV, even the written word—to understand that every pivotal communication medium has gone through a similar cycle of euphoria and hysteria before approaching balance.

The response to the global pandemic and the extraordinary impact of the Black Lives Matter movement has only deepened my conviction that social media is a foundational technology—one that's finding its footing and will only prove more transformational in years ahead.

For all its flaws, social media has fundamentally reshaped how we live, think, work, and experience life. It's allowed us to connect, to organize, and to keep businesses open in unprecedented times. It's also allowed us to question the status quo, hold institutions and government accountable, and force society to do better. It's the soil that our postpandemic world will grow from, and it's up to us to keep it healthy, to seed it with the right intentions and cultivate an environment that will protect its potential and allow it to thrive.

There's no escaping this mass cultural shift. The pendulum has swung: social media is intimately integrated into the fabric of our lives and jobs, and the responsibility now hangs on all of our shoulders to understand and use it right. We've reached a critical tipping point in our digital journey of connection where it's become a hazard to not ask the tough questions, embrace new opportunities, and find the right path forward. I hope that this book can serve as a guide for those looking to understand both the potential and the pitfalls of social media and map out a plan to thrive, as businesses and as people, in this new landscape in the years ahead.

PART 1

SOCIAL WORLD

1

Understanding the Tech Life Cycle— Euphoria, Hysteria, Balance

With a new medium, it starts with euphoria and then goes to hysteria and then hopefully you get some kind of balance. It happened with the radio. This happened with TV. There was a huge amount of skepticism about reading Plato because he was writing and no one could argue versus yelling into a public square.

—ADAM MOSSERI, CEO OF INSTAGRAM

Guglielmo Marconi was a trendsetter, in more ways than one.

He was a snazzy dresser who favored crisp shirts with rounded collars and immaculately tied neckties. You can't find a photo of him in which a strand of hair is out of place. And then there was that whole radio thing.

Born in 1874, the Italian inventor was one of the original disruptors, somebody who began his quest to change the world at the tender age of twenty. Fascinated by the work of radio wave pioneer Heinrich Hertz—that's *Hertz*, as in *megahertz* and *kilohertz*—young Marconi planned, plotted, and executed the invention of the radio transmitter from his attic.

Relatively speaking, Marconi was a fast worker. By 1901, the young man's radio transmitter was able to send a signal a whopping 2,200 miles. (For context, that's the distance between New York City and Costa Rica.) Think about that: in 1894, the kid had hertz and a dream; seven years later, he altered the manner in which human beings communicate with one another. And he did it all without a series A.

In the United States, regulated radio became a thing

several decades later on September 15, 1921, when WBZ in Springfield, Massachusetts, started broadcasting. Over the next several months, WBZ delivered political speeches, opera performances, and even a bilingual show from Montreal.

But it wasn't long before unexpected challenges surfaced. One of the first instances of radio getting real came about in 1924, when Pentecostal evangelist and founder of the International Church of the Foursquare Gospel Aimee Semple McPherson purchased her own station, Los Angeles's KFSG, so she could discuss her *individualistic* views on politics and religion—without any kind of oversight. Like this bit:

> And my task, as I see it, is to interest you folks to help me, to help them, to join the line right around the whole world! Not only to help the heathen abroad, but to help the heathen in Los Angeles. In America, too.

Seems like Aimee believed the country was overflowing with heathens, and a goodly number of people with listening range of KFSG agreed. Soon, the airwaves were clogged with righteous evangelicals and their secular adversaries feuding back and forth, like the twentieth-century equivalent of Twitter trolls. Ordinary listeners, initially enchanted with the new technology, were suddenly caught in the middle. But rather than just turn the radio dial, they did something. Fed up with all the bickering, radio's early power users raised a stink. And the government listened.

Three years after Aimee's show premiered, President Calvin Coolidge signed the Radio Act of 1927, which regulated "all forms of interstate and foreign radio transmission and communications within the United States, its Territories and possessions." The act went on to say that radio stations had to deliver content that was "in the public interest, convenience, or necessity." An overseeing body called the Federal Radio Commission was soon created.

Realize, this wasn't about obliterating anybody's right to free speech. Broadcasters were still allowed to discuss religion and politics, but content had to serve a public good. In time, radio went from being the Wild West of media—a free-for-all in which broadcasters could say or do most anything, no matter how offensive—to a well-regulated entity that will likely be with us in one form or another forever.

So, to summarize: We're looking at a new paradigm-shifting form of communication technology that changed the way people interacted with one another, impacted how they received their news, and created an entirely new ecosystem. Initially, listeners were entranced and entertained and felt nothing could go wrong. Then, while the medium was still in its infancy, a vocal minority of individuals with worldviews that didn't align with those of the majority ruffled some feathers, oftentimes overshadowing all the good that the technology had accomplished. Concerned citizens and eventually government regulators stepped in, gradually striking a balance between personal freedoms, business needs, and public good.

Sound familiar?

***

In terms of sartorial slickness, Facebook founder Mark Zuckerberg isn't necessarily Guglielmo Marconi.

His hairstyle is nondescript, and unless he's speaking to Congress while looking uncomfortable in his suit,

Zuckerberg's attire generally consists of jeans and a T-shirt. As a disruptor, however, Zuckerberg could give Marconi a run for his money.

Chances are, you know the broad strokes of Facebook's origin story: Harvard computer genius creates social network in his dorm room; social network attracts millions upon millions of happy users; company's valuation skyrockets; bad actors start manipulating user data and company algorithms; Russian hackers use the platform to spread misinformation in swing states during the 2016 US presidential election—i.e. Cambridge Analytica; Facebook's reputation is impacted and users begin to lash out; adjustments are made internally, while scrutiny is applied externally by Congress and public interest groups … aaaaaand here we are.

Thanks to these bumps in the road—bumps that virtually all new technologies experience in one manner or another—the good will get better.

Just as was the case with radio, social media has had its share of bumps in the road. And thanks to these bumps in the road—bumps that virtually all new technologies experience in one manner or another—the good will get better. Yes the handling of user data needs to be addressed, and the issue of political content on social platforms remains a lightning rod. But remember, social media still boasts all of the factors that have attracted more than half of the world's population as users. You can connect with friends and family, shop your favorite brands, or order food delivery when the world is shut down. You can create or join a coalition of like-minded altruists and spark a global movement in response to racial injustice. More than that, though, new applications for social media are emerging and spreading at an ever-accelerating pace—from ways to browse and buy products on Instagram to tools for letting "virtual fans" join sporting events from home. As a vehicle for customer service, commerce, and genuine human connection, this is just the start.

To be sure, we're still in the early stages of adjustment, input, and evolution. It's critical to remember that less than twenty years ago, social media didn't even exist as a communication option. The good news is public scrutiny over the lack of transparency and protection of user data is not falling on deaf ears. Solutions to key challenges such as misinformation, propaganda, and bullying are being discussed and explored by business, government, and the networks themselves. Meanwhile, government, consumers, and tech leaders are starting to raise questions and propose solutions in an attempt to balance free speech, public good, and business necessity.

The process isn't easy. It involves competing interests and very different priorities. It requires reconciling the desire for scale and efficiency with the need for accuracy and social responsibility. But, as is the case with so many technologies, from radio to rockets, it has to be done. And the good news is that we've started that journey.

In the end, social media remains the most robust marketplace of ideas the world has even seen—ungated, infinitely rich, and with truly global reach. If we're ever going to have the hard conversations that are so desperately needed—on inequality, on environment, on our collective future—then social media may just be the best platform to make that happen.

And as a technology, it's not going anywhere. The rise of new networks and technologies, from TikTok to audio-first network Clubhouse, illustrates that while platforms may come and go, the medium is here to stay. For businesses, social media will only be an increasingly powerful way to engage stakeholders in the years ahead.

But reaching employees, consumers, investors, and the public depends on acknowledging the evolution of social media as a technology and the shifting expectations that go with it. Companies that understand the new balance of social power stand to thrive. Those that don't may well go the way of the telegraph.

And to understand the new social paradigm, we need look no further than Greta Thunberg.

CHAPTER 2

The Rise of
Digital Natives

reta Thunberg is anything but an extrovert. In fact, not long before the sixteen-year-old was lauded with Nobel Peace Prize nominations and *TIME* magazine's most prestigious cover, she secluded, having fallen into a deep depression. Greta's mother, Malena Ernman, recounted how her daughter "was slowly disappearing into some kind of darkness," in her book coauthored with Greta and her other daughter, Beata, *Our House Is on Fire*.[1]

---

1   Malena Ernman, Greta Thunberg, and Beata Ernman,
    Our House is on Fire: Scenes of a Family and a Planet in Crisis
    (London: Penguin, 2020).

Malena wrote, "She stopped playing the piano. She stopped laughing. She stopped talking. And she stopped eating." Gravely concerned for their daughter's well-being, Greta's parents sought out a doctor's evaluation. Greta was diagnosed with Asperger's syndrome and obsessive-compulsive disorder. Changes were made to her routine and school environment. Greta slowly started eating again, and life leveled out. However, her depression and the radical impact it had on her behavior couldn't be neatly explained by a psychiatric label.

One day, a film was shown in Greta's class about garbage in the South Pacific. Like many of her classmates, Greta was gripped with concern; but unlike her classmates, who quickly moved on to conversations about travel, shopping, and the distractions of their iPhones, Greta couldn't reconcile what she had learned about climate change with the half-hearted efforts the leaders of society seemed to be making to rectify it.

On August 20, 2018, Greta woke up early. She packed an extra sweater and cushion in her backpack and set out on her bike not to her first class but to Parliament in her home city of Stockholm, Sweden.

Armed with one hundred flyers that had facts about

climate change and sustainability, she set down a hand-painted sign made of scrap wood that read School Strike for Climate.

Her father took a photo of her on her mobile phone. A passerby took another. Greta posted the pictures on social media. She had fewer than twenty followers on Instagram and a few more than that on Twitter, but that was all about to change.

Shortly after Greta posted her photos, Staffan Lindberg, a political scientist from Stockholm, retweeted her post. Meteorologist Pår Homgren followed suit, then singer-songwriter Stefan Sunstrom. The retweets and shares accelerated quickly.

Twelve months later to the day, four million people joined Greta for a global climate strike, the largest climate demonstration in human history. Greta went from obscurity to addressing the heads of state at the UN, meeting the pope, and even standing up to the president of the United States. In sixteen months, her social media following grew to exceed Greenpeace's membership of three million, a membership that took nearly fifty years to amass. But unlike Greenpeace, Greta's not an international group. She's just one person, one teenager who happens to be a digital native.

Greta's generation has grown up with social media and smartphones. It's fully integrated into their young lives, just as TV and landlines and even radio were for generations before. Digital formats are their preferred method of communicating. It's not right. It's not wrong. It just *is*. Never having navigated life without social and digital media, Greta and her fellow digital natives have been transformative in terms of how they've mobilized people, how they've communicated, and how they've gotten *out there*. She created a movement, with unprecedented speed and at unprecedented scale, that wouldn't have been possible without social media.

We've entered an era where social natives have more economic capital than ever before.

In many ways, Greta's story is illustrative of the power held by the upcoming generation. We've entered an era where social natives have more economic capital than ever before. Gen Z will account for 40 percent of global consumers this year, with spending power of nearly $150 billion.[2] And these digitally savvy eight- to twenty-four-year-olds are educating the generations that came before them on new, efficient ways to communicate, shop, and find and share information. Ultimately, they're reinventing the rules for business because their expectations are entirely different.

Like Greta, this generation is less interested in fitting in and more interested in finding their tribe, being born with the tools to do just that. Rather than forcing themselves into a popular identity, they've grown up exchanging Lego ideas with fellow Eurobricks on TikTok or sharing fan art over Snapchat with peers in Singapore. Licensed to explore all facets of their identity, they have the freedom to define themselves authentically and, through online communities, find acceptance more easily. Having grown up online, they also experiment with platforms in ways that suit their own specific needs. They're less likely to use social media the way it was intended and more likely to manipulate functions

---

2   Imran Amed, Anita Balchandani, Marco Beltrami, Achim Berg, Saskia Hedrich, and Felix Rölkens, "The Influence of 'Woke' Consumers on Fashion," McKinsey & Company, February 12, 2019, https://www.mckinsey.com/industries/retail/our-insights/the-influence-of-woke-consumers-on-fashion#.

and features as a form of self-expression or way of connecting with others.

This inherited right to expression and intrinsic need to find their "true" selves plays into how they interact with brands. Gen Z expects brands to have an authentic and transparent digital identity. Social media is where they both establish and share their core values and beliefs, and they expect brands to do the same. Nearly 70 percent of Gen Zers expect brands to contribute to society, and an overwhelming majority are willing to pay more for products or services that are produced in an ethical and sustainable way, according to global consulting firm McKinsey & Company.[3]

In many ways, how Gen Z utilizes the digital landscape represents a fundamental but often overlooked shift in the application of social media. For them (and generations going forward) social media is anything but a marketing channel—it's a vehicle for self-expression and finding community. Chastened by privacy abuses, trolling, and

---

3   Tracy Francis and Fernanda Hoefel, "'True Gen': Generation Z and Its implications for Companies," McKinsey & Company, November 12, 2018, https://www.mckinsey.com/industries/consumer-packaged-goods/our-insights/true-gen-generation-z-and-its-implications-for-companies.

fake news, they're determined to find a different way to use social networks. This technology for them still has tremendous value—but it's all in how it's used.

To analogize using another technology, ask yourself, Is fire bad? It can be. It could burn down your house. It can disfigure you. But it also cooks your food, and on a cold night, there's no doubt about the value fire brings.

Which brings us back to Greta, because Greta uses this new digital fire at its best. She's what's *great* about social media.

Whether by design or intuition, this young woman harnessed the power of social media and is bringing about positive change in the world as a result. Case in point, in 2019, Sweden has reported a 4 percent drop in domestic air travel, as well as an increase in transportation via rail, thanks in major part to her awareness raising (and this was before COVID-19 shut down so many airlines).[4]

Her impact has been so great that she has *Fortune* 500 executives asking themselves, "What has my company

---

4 "Sweden Sees Rare Fall in Air passengers, as Flight-Shaming Takes Off," BBC News, January 10, 2020, https://www.bbc.com/news/world-europe-51067440; "Sweden's Rail Travel Jumps with Some Help from 'Flight Shaming,'" Reuters, February 13, 2020, https://reut.rs/2SN4WKe.

been doing to make certain we're carbon-neutral by 2035?" Greta and her four million Twitter followers have impacted the thought processes of adults who have been in the corporate world for decades. So what does that mean for business?

Here's what: in today's business world, our scope of concern has to extend beyond just customers to embrace the community at large. The era of catering just to *shareholders* is coming to a close. Increasingly, businesses need to show consideration for *stakeholders* of all kinds, a group that embraces everyone from employees and suppliers to marginalized people and the general public. In the social media era, driven by the power and passions of Gen Z, this is no longer a choice; it's an obligation. It takes thirty seconds for somebody to create a tweet that goes viral, and that tweet can impact share prices within minutes. The transparency ushered in by social media has blurred lines between shareholders and stakeholders and requires businesses (at least ones that want to thrive) to balance economic growth with community interest.

The Greta example also shows us that social media, for all its growing pains, is undoubtedly here to stay. As such, finding ways to forge relationships with consumers via social channels will be one of—if not *the*—primary mandates of tomorrow's business leaders.

The Black Lives Matter movement presents another example both of the power of social media and the need for brands to engage with transparency and authenticity.

Beginning in July 2013, when the hashtag #BlackLivesMatter spread in response to the acquittal of George Zimmerman, who shot and killed teenager Trayvon Martin in February of 2012, Black Lives Matter has evolved into an international movement that continues to gain momentum. In 2020, Black Lives Matter protests took place around the country in reaction to the killing of George Floyd by a police officer, and consumers leaned on the movement's social media origins to drive brands and organizations to acknowledge BLM and to take a stance. For businesses, communicating clear values—and showing awareness of all stakeholders, not just customers—has become a business (and moral) imperative. A brand's social media is one of the key ways to communicate these beliefs, values, and goals, and it's where many consumers will turn first for answers—and to complain if the brand's response is inadequate.

This is a perfect demonstration of global connectedness changing how sellers interact with consumers and how consumers interact with one another. This is how the digital natives do it—it's what all of us are adapting to do—and we're not going back to the old paradigm. Digital natives are sophisticated in terms of what they know, what they want, and how they want it. Even more importantly, they know how to use social media, and they know if *you* know how to use social media. If you speak down to them, if you don't treat them with the respect they deserve, if you do the wrong thing, you lose. And in today's business climate, it's all about doing the right thing.

If you don't believe me, check out Greta.

PART II

SOCIAL
BUSINESS

CHAPTER 3

Trust through Transparency

I n early February, I was doing something that now seems highly unusual: having a meeting in my office. Members of the communications team had gathered around for a weekly meeting on current events in the social media world, from the latest memes going viral to efforts to crack down on fake Twitter accounts.

But on this day, all the talk was about a novel virus that had quickly swept through China and was starting to take lives in parts of Europe. From our office in Vancouver, Canada, it was like watching a distant tsunami approach, and all the speculation was around when it would hit North America and how hard.

Over the following weeks, we saw whole populations confined to their homes. We saw doctors and nurses—who'd become frontline workers—asking people to socially distance to flatten the curve and avoid the devastation in places like Italy, where one doctor described how his hospital had become a "war zone." As a global company, one by one our offices in affected regions moved to remote work. Within a matter of weeks, the whole world had taken shelter. Businesses were forced to shut down overnight or pivot to new models. Amid all the chaos of what quickly escalated to a global crisis, social media became a lifeline.

And a whole lot of people utilized that lifeline. In the days and weeks following the global lockdown, our platform alone saw a 15 to 20 percent increase in posts from our eighteen million users, as companies reached out to customers and employees. A study of twenty-five thousand consumers across thirty markets showed engagement increased 61 percent over normal usage rates.[5] Messaging across Facebook, Instagram, and

---

5   "COVID-19 Barometer: Consumer Attitudes, Media Habits and Expectations," Kantar.com, April 3, 2020, https://www.kantar.com/Inspiration/Coronavirus/COVID-19-Barometer-Consumer-attitudes-media-habits-and-expectations.

WhatsApp increased 50 percent in countries hardest hit by the virus, and Twitter saw 23 percent more daily users than in the year previous.[6]

Not only was the population craving news but also something just as important in a time of social distancing—connection.

Within days of the government-enforced lockdowns, we saw Jacinda Ardern, the prime minister of New Zealand, take to Facebook Live wearing a worn sweatshirt after tucking her toddler into bed to field questions from her people. We saw chefs, forced to close down restaurants overnight, open up their home kitchens to offer cooking lessons on Instagram. And in our news feeds, we saw healthcare workers donning protective masks and gloves holding up signs of gratitude for the 7:00 p.m. cheer that for many became the most social part of an otherwise isolated day.

On our own platform, we saw businesses and consumers alike taking to social media as they never had. Marketing and ads gave way to direct engagement—

---

6   Alex Schultz, "Keeping Our Services Stable and Reliable During the COVID-19 Outbreak," Facebook Newsroom, March 24, 2020, https://about.fb.com/news/2020/03/keeping-our-apps-stable-during-covid-19/.

one-on-one interaction with other people in an effort to stay connected in a world that had otherwise all but shut down.

And with so many businesses forced to close (at least temporarily) and/or lay off employees, brands suddenly realized the importance of connections over conversions. They quickly turned to social media to find new ways to engage with customers. And the ones that used social media to its fullest—to reach out to customers, reassure employees, and support their community—shared a key attribute.

Purpose.

> In the digital age, it's more imperative than ever that brands remain clear on their purpose—and not just in *what* they do but *why* and *how* they do it.

In the digital age, it's more imperative than ever that brands remain clear on their purpose—and not just in *what* they do but *why* and *how* they do it. Businesses that have a clear purpose and operate in line with their values stand out. They find creative ways to care for their employees and consumers, earning trust and loyalty during a critical moment in history.

And never was there a greater test for brands to

operate in line with their values than when the world was forced to a halt due to COVID-19. Countless companies stepped up to the occasion. From our own customer roster at Hootsuite, there were so many inspiring examples.

Avon, a company with the public mission to improve the lives of women globally, quickly found a way to support its community when reality shifted. Noting 1.6 million women had experienced domestic abuse in the last year, the company took to social media to connect with women in self-isolation, sharing a commitment to keeping frontline services open for women facing domestic violence while donating $130,000 to Refuge, a national domestic abuse charity.

Melia Hotels, a family company committed to contributing to a better world, utilized social to communicate and organize as it turned eleven of its hotels in Spain into temporary hospitals.

And adult beverage brands like Bacardi and Jim Beam, who pride themselves on producing premium products, showed good taste in pivoting to produce gallons of hand sanitizer to protect against the virus while sharing info and updates on social media.

But it's not enough to have purpose. Expectations

around transparency have vastly increased, forcing brands to not only share their values but to truly live them. Authenticity is the currency of social media. Brands who don't practice what they preach risk irreparable damage, especially in an era when an unsavory smartphone video can go viral in a matter of minutes.

Businesses that master the art of both purpose and transparency on social media build a strong community of on- and offline consumer advocates who will champion their brand and help mitigate risk when things go south. Because let's face it: even the best brands don't always get it right.

Tesla's and Space X's Elon Musk is known for his social media acumen. But he miscalibrated—big time— on March 6, 2020, when he tweeted, "The coronavirus panic is dumb." The tone deafness continued later that week, when he tweeted about what he believed to be a "fatality rate [that is] also greatly overstated," as well as a theory that the stock market "was a bit high anyway" and was "due for a course correction."

Not surprisingly, the reaction was swift and severe. This response from @LM2020Texas summed up the vibe:

LMTexas
@LM2020Texas

I'm the biggest Elon Musk fan and he is so wrong to tweet this, he is probably doing it because he doesn't like how the virus has negatively affected his company revenues etc he's already successful but guess more money is always more important than lives to some people

Musk noticed the backlash—how could he not? Then he did something as refreshing as it was socially savvy: he admitted that he had been dead wrong, and he put his money where his mouth was.

On March 18 and 19, Musk exchanged tweets with New York City mayor Bill de Blasio. He pledged to use Tesla's and Space X's facilities and manpower to manufacture ventilators for New York, the American city hit hardest by the coronavirus.

What's critical here—and what makes Musk such a paradigm of effective social media leadership—is that he changed course and tuned into his fan base on social media for suggestions on how to help. True to a brand built on bold promises, he then took swift action, committing to producing ventilators. Through it all, there was a transparency—a real-time window into Musk's values, his intentions, his soul-searching—that's the real cornerstone of effective social media. The consequences of inauthenticity extend far beyond bad optics. How a company is perceived now, more than ever before, impacts its bottom line. Back in the day, if you were turned off by a brand or a store, you'd simply stop patronizing them. Today, consumers can take action with a few taps and clicks. One form this takes is the "buycott," meaning they'll turn to your competitor and encourage everyone within their network to do the same.

For a pandemic-related example with a less happy resolution, let's look at Amazon. Jeff Bezos got spanked on social when employees at Amazon and Whole Foods—both companies staffed with "essential workers" expected to keep working during the crisis—weren't given proper protection from the virus: no mandatory social distancing, no extended sick leave, little paid time off. A tweet from user @MaryJaneUSA1 in May of 2020—just three months after most states locked down—said:

MaryJaneUSA1
@MaryJaneUSA1

Bezos got $35 BILLION richer in just 10 weeks but is dropping $2 hr. hazard pay WHILE AMAZON PEOPLE KEEP CATCHING CORONAVIRUS! How greedy can Bezos be?? It's outrageous to treat employees so badly who generate the $$.

Bezos went on to make a $100 million donation to Feeding America, a nonprofit that distributes money to food banks across the United States. It didn't satisfy critics concerned over conditions in his own businesses, especially after it was reported on April 15 that due in part to Amazon's revenue growth during the crisis, Bezos's net worth jumped an additional $23.6 billion.[7]

Financially speaking, Amazon and Whole Foods will be fine in the short term, but their reputation has been tainted, and that could become problematic when the business world gets back into full swing. That might not be the case when they again have broader choices.

---

7   Isobel Asher Hamilton, "Jeff Bezos Is Wealthier by $24 Billion in 2020, as Amazon Reports at Least 74 COVID-19 US Warehouse Cases and Its First Death," Business Insider, April 15, 2020, https://www.businessinsider.com/jeff-bezos-net-worth-jumps-23-billion-during-coronavirus-crisis-2020-4.

The reality, as Edelman noted in its latest Trust Barometer report, is that "people today grant their trust based on two distinct attributes: competence (delivering on promises) and ethical behavior (doing the right thing and working to improve society)." Social media is now the prevailing platform for expressing, exploring and vetting company values. This is why you, as a company, need to be transparent in your purpose. Businesses who fail to appreciate that shift may not remain in business for long. Those who do "get it," however, will create enduring loyalty and be in it for the long haul.

Transparency does have limits, of course. No company can be transparent about all facets of their operations—there's privacy to keep in mind, not to mention strategy and competitive advantage.

But if a company does make a mistake or faces a challenge that could be potentially damaging to its reputation, it pays to be open and honest. By telling the story themselves, companies can sometimes get ahead of detrimental information and may also be able to strengthen their connections with stakeholders or consumers by showing the courage to expose their less flattering aspects. Companies will increasingly have to understand this dynamic in order to survive in a world where nothing can be hidden.

Transparency around a brand's purpose builds trust and community, plus it shines a light on competitor differentiation, aids in risk mitigation, and strengthens the employer's brand. Today's consumers expect to be able to learn about a company beyond product information and want to see diversity and values reflected in the way they operate. And social media is the gasoline accelerating this fire, as brands and consumers have unprecedented access to one another.

PRO TIPS

1. Own your narrative before the media or another company hijacks it.

2. Know your position and your tribe. Don't try to be all things to all people. Find your audience, and make them yours.

3. Walk the talk. Social media has greatly expanded our ability to share and access information, which has sharpened consumers' bullshit detectors. This has resulted in backlash for brands who don't say what they do, then do what they say.

4. Empower customers with data and information. Consumers are smarter than ever, and if you don't treat them as equals, they'll go elsewhere—in a heartbeat.

5. Fess up when you get it wrong. Everybody makes mistakes. Even Elon Musk.

CHAPTER 4

From the Town Square to the Living Room

S ocial media changed forever on the morning of March 6, 2019, when Mark Zuckerberg—who built Facebook's empire on the premise of free and open sharing—published a blog post entitled "A Privacy-Focused Vision for Social Networking." (Google it. It's a great read.) In it, he wrote:

> I believe the future of communication will increasingly shift to private, encrypted services where people can be confident what they say to each other stays secure and their messages and content won't stick around forever. This is the future I hope we will help bring about.

We plan to build this the way we've developed WhatsApp: focus on the most fundamental and private use case—messaging—make it as secure as possible, and then build more ways for people to interact on top of that, including calls, video chats, groups, stories, businesses, payments, commerce, and ultimately a platform for many other kinds of private services.

Zuckerberg went on to list the guiding principles behind his plan: private interaction, encryption, reducing permanence, safety, interoperability, and secure data storage.

All this adds up to a blunt concession: the idea of social media as a public broadcast channel as a way for anyone to reach a mass audience, build a following, and change the world needed updating. The platform envisioned when drawing up Facebook's initial game plan had been compromised by cybertrolls, bullies, invasive ads, Russian interference, and addictive algorithms.

In short, it's time for a pivot. The new paradigm emphasizes places for people to connect either one-on-one or within small, closed groups. In this updated social landscape, the town square vibe is giving way to a living room. Privacy, authenticity, and security—embodied in Facebook's growing empire of messaging apps— represent the new order of the day.

> The new paradigm emphasizes places for people to connect either one-on-one or within small, closed groups. In this updated social landscape, the town square vibe is giving way to a living room.

A logical approach, certainly, but it raises the question, where does this leave brands that have come to rely on social media as a primary means of reaching their audiences? How can companies that depend on social to engage customers, employees, and other stakeholders adapt and keep up as the platforms change? And can users accept this new direction?

Facebook's Stories feature is certainly easing the transition.

Back in 2013, Snapchat debuted Stories, a collection of vertical, ephemeral slideshows that comprised a mix of photos and videos shot by users over the course of a day. Stories was introduced to a much wider audience on Instagram in 2016, and WhatsApp rolled it out the following year.

Initially, Stories didn't make much of a dent in the social world, but today they're quietly reshaping it, fundamentally altering how we share and consume content. For companies that rely on social media to reach their customers, this presents brand-new opportunities—and brand-new challenges.

It's difficult to ignore the power—and potential ROI—of Stories. In fact, Facebook's own chief product officer, Chris Cox, has pretty much hitched the company wagon to Stories, noting in 2018, "The Stories format is on a path to surpass feeds as the primary way people share things with their friends sometime next year."[8] In other words, embracing Stories is no longer an option for businesses but rather a requirement.

And digital natives are totally cool with that.

This group has grown up saturated with digital marketing and content. They've learned to tune out banner ads and can smell a sales pitch a mile away.

---

8   Josh Constine, "Stories Are About to Surpass Feed Sharing. Now What?," TechCrunch.com, May 2, 2018, https://techcrunch.com/2018/05/02/stories-are-about-to-surpass-feed-sharing-now-what/.

Companies hoping to reach them with Stories need to do so by providing true value. They have to entertain, inform, or educate ... *and not just sell.* Far from a direct marketing or sales play, Stories are a branding opportunity, with little place for a heavy-handed call to action.

Stories work best when they integrate video, text, and images. Though they might look "off the cuff," Stories often have higher production value and require greater technical expertise than a typical tweet or Facebook post. As noted by TechCrunch's Josh Constine, "Advertisers must rethink their message not as a headline, body text, and link, but as a background, overlays, and a feeling that lingers even if viewers don't click through."[9] It's a whole new way of telling a story, and if you want your business to thrive, you need to figure it out ... or hire an agency that already has.

Orange juice giant Tropicana immediately recognized the potential of Instagram Stories. In one campaign, it combined mouthwatering pour shots of juice being mixed into festive drinks like sangria. Hand-drawn text and arrows offered mixing instructions, and users were

---

9   Ibid.

invited to "swipe up" for the full recipe. The result: An eighteen-point lift in ad recall and measurable boost in purchase intent.[10]

Today, the traditional pack shot—a sterile image of a product sealed tightly in its packaging—has little place in the realm of Stories. (Understandable. When's the last time you looked at a pack shot and thought, "I'm buying *that*"?) Successful brands are instead using the multimedia format to show how products fit into the context of customers' lives. Tapping into influencers to create and share product, Stories enables companies to extend their reach and access an already bought-in audience.

Interestingly, however, fancy graphics and slick editing aren't as important to Story success as something far more valued by digital natives: authenticity. After wrapping up an Instagram Stories campaign, the *Guardian* made a critical discovery: highly scripted stories were not providing the expected return on investment, while their lo-fi Stories—which featured casual language and emojis—performed beyond anybody's

---

10   Ryan Holmes, "4 Tips on How to Succeed with Stories," *Fast Company*, October 22, 2018, https://www.fastcompany.com/90252767/4-tips-on-how-to-succeed-with-stories.

wildest dreams. On the strength of these DIY-looking stories, the *Guardian* grew their Instagram followers from 860,000 to 1 million in just four months.[11]

You can't talk about Facebook Stories and ephemeral content without talking about another key social media innovation: Facebook Groups. Groups—dedicated micro-communities for particular interests, from *Game of Thrones* to Peloton cycles—truly tie together Zuckerberg's rebooted vision. Plus, Facebook Groups can be awesome. Just ask Condé Nast.

A couple years back, *Condé Nast Traveler* did something a little unusual in the social media universe: it played hard to get. Instead of courting new followers with clickbait and promo codes, the company required that interested users apply to get into its closed Facebook Group, which focused solely on female travelers. To be considered for membership, applicants had to explain why the group was important to them and show an understanding of the community guidelines. That led to fifty thousand members and a level of activity many

---

11 Jessica Davies, "The *Guardian* finds less polished video works better on Instagram Stories," *Digiday*, July 5, 2018, https://digiday.com/media/guardian-finds-less-polished-video-works-better-instagram/.

brands could only dream of—three-quarters of users were active on a daily basis.[12] The initiative was so successful that Condé Nast has since extended Facebook Groups across eight of its brands, among them the *New Yorker*, *Vanity Fair*, and *Allure*.

The safety of the group (not to mention the general lack of snark) allows brands to interact and connect with their users on a human level—less transaction and more actual communication. In a group, a business can nurture intimacy, build trust, reward loyalists, and, yes, create a community.

A group can also act as the biggest focus group you've ever had. If your company's group has proven to be a quality, trustworthy, intelligent batch of people—if they don't populate your feed with noise—go ahead and float an idea past them. Do you like logo A or logo B? Would you prefer pumpkin pie, chocolate mint, or blueberry? If we're making merch, would you rather see a T-shirt or a hoodie? You'll get dozens, if not hundreds of fast, unedited, definitive responses, the kind of thing you'd

---

12   Ryan Holmes, "Are Facebook Groups the Future of Social Media (or a Dead End)?," *Forbes*, October 29, 2018, https://www.forbes.com/sites/ryanholmes/2018/10/29/are-facebook-groups-the-future-of-social-media-or-a-dead-end/#4551b73e1d23.

never get out of a traditional focus group. At Hootsuite, we've utilized Facebook Groups for these exact reasons.

Now, let's step away from the town square and the living room and head "behind closed doors," to the realm of messaging. Messaging in the social media context refers to one-on-one, typically private communications: chats but handled by social platforms.

One critical use case for messaging for companies is in the realm of customer support. We can all agree that forcing customers to endure an endless phone menu and then sit on hold for twenty minutes or having them wait eight hours for an email response is not the most efficient manner in which to service their needs.

Messaging offers a powerful alternative. It's a real-time way for customers to get relief and provides a permanent record of interaction, all compressed into a tidy dialogue. In contrast to social media, messaging is a private, one-to-one experience, with rants and sensitive personal info kept well out of the public domain. Additionally, photos, videos, and audio can all be effortlessly incorporated, a boon for documenting problems or explaining complex issues.

For all the talk of "omnichannel" customer service,

however, businesses have been slow to embrace messaging. Only a small fraction of all companies on Facebook currently use Messenger each month to talk to customers. But those turning to messaging have seen compelling results. KLM became the first airline to expand its service to WhatsApp, allowing customers to receive flight status updates and get 24-7 service in multiple languages. H&M gives fashion advice through Kik, while Domino's helps customers find coupons and make delivery orders via Facebook Messenger.

Admittedly, the shift to messaging and its always-on availability can be jarring for businesses, and I saw this firsthand when Hootsuite rolled out some new functionality to our own platform. Our users loved the new feature, but it also led to a record number of support requests via social, chat, email, and other channels—75,000 in one quarter alone. We were swamped with questions on functionality, setup, and how-tos for months.

Keeping up with this messaging revolution requires a ground-up rethink about how we do customer service, specifically the question of scale. When customers have more ways to reach out than ever before—and can do so at a moment's notice—how do you respond in a way that's both meaningful and timely?

One answer: Chat bots. I know, I know: bots have gotten off to a bumpy start. Facebook's first-gen bots reportedly failed to understand users 70 percent of the time, and even so, eight of ten consumers prefer to interact with a real person.[13] But that's largely because too many companies are trying to build a human in digital code, pushing today's tech past its limits in the name of cost cutting.

Instead, the focus for now is best narrowed to addressing the most common customer queries with bots, the questions that come up time and time again. At Hootsuite, the same twenty questions—mostly stuff about billing and setup—account for 80 percent of our customer inquiries. Handling these simple, repetitive issues are what bots were made for, freeing up human capital for more complex questions and needs.

But that's only part of the story. While bots represent the front line in this brave new era of messenger-based customer service, equally critical is how efficiently—and

13 Andrew Orlowski, "Facebook Scales Back AI Flagship after chatbots Hit 70% Failure Rate," the Register, February 22, 2017, https://www.theregister.com/2017/02/22/facebook_ai_fail/.

intelligently—customers are escalated to human agents. In fact, this is what makes or breaks the whole experience. Makeup company Sephora offers a model for how to do this right, deploying an army of chat bots to help with tutorials and product suggestions but having real humans on call when a customer requires more help.

AI is proving just as critical behind the scenes. As messaging volume swells, smart tools are increasingly needed to automatically tag and route messages—whether they come in via social media, messaging apps, or another channel—for human follow-up. This ensures the right information quickly gets to the right person inside a company.

What's clear above all is that messaging is becoming *the* dominant paradigm for customer engagement. WhatsApp is up to 1.5 billion users in 180 countries.[14] Facebook is hoping to unite all the apps in its messaging kingdom—a kingdom that houses some 2.6 billion users—enabling seamless cross-platform pinging across Instagram, Facebook, and WhatsApp.

---

14   Mansoor Iqbal, "WhatsApp Revenue and Usage Statistics (2020)," Business of Apps, updated June 23, 2020, https://www.businessofapps.com/data/whatsapp-statistics/.

Stories, groups, and messaging may seem like disparate elements in the social media ecosystem, but thanks to Zuckerberg's new *new* vision, they'll be tied together in a neat little bow. The business and social media worlds will be better for it.

And if nothing else, it'll make the living room that much more interesting.

PRO TIPS

- Strike a balance with stories. Users expect a certain degree of polish from brands, but too much editing can rob a story of its authenticity. Remember that attention spans on social media are measured in seconds and the rise of ephemeral content means it disappears like *that*.

- Move fast. If you're not already leveraging messaging, groups, and ephemeral updates, you're losing out on valuable touch points. Tech advances and generational trends are making these channels more powerful than ever.

- Be where your customers are. Consumers expect to be able to connect with brands where, when, and how they want. If you're not catering to their individual preferences, you risk losing market share to competitors who understand the value of convenience and accessibility.

- Allow for private conversations. Giving customers easy access to secure, private channels to communicate with your brand not only deescalates challenging situations, but it also allows for the exchange of personal information as needed.

CHAPTER 5

Advertising and Influence

oday, what greases the wheel of social media is paid advertising.

The days of just "going viral" organically are long gone, and network algorithms severely curtail how many of your own followers you can reach—unless you're willing to spend to extend that reach. Put simply: To maximize your reach and impact, paid campaigns *have* to be a part of your strategy on social media.

There's just one problem.

With the constant evolution of the social space, if you and your team don't understand what your customer acquisition costs are and how advertising impacts your returns, *you shouldn't be doing it.*

Just take the example of Procter & Gamble.

In 2016, P&G spent $7.2 billion on advertising, the most of any company in the world. A year later, however, the consumer goods giant decided to cut its digital ad spend—a third of its entire budget—by $200 million. And an interesting thing happened: P&G increased its reach by 10 percent.[15]

P&G discovered not only that their ad spend wasn't reaching their target audiences effectively but that how consumers were viewing ads online was different. For instance, the data showed the average view time for an ad on a mobile phone was a mere 1.7 seconds. Reaching audiences on digital would require new, creative approaches.

Think about that. If one of the biggest companies on the planet—with help from world-class media-buying agencies—couldn't throw money at social media and succeed, then no one can.

So what's the right way to incorporate social media

---

15   Lucy Handley, "We Need 'Fewer' Ads, Says Consumer Goods Company That Spent $7.2 Billion on Advertising in 2016," CNBC, updated April 13, 2017, https://www.cnbc.com/2017/04/05/we-need-fewer-ads-says-pg-which-spent-7-2bn-on-adverts-in-2016.html.

ads? For starters, make sure you know who you want to reach and where they live online. Looking to access influential business professionals? LinkedIn is the place to be, and they conveniently offer robust ad tools. Going after millennials and Gen Zers? Set your sights on Instagram, which uses Facebook's unrivaled ad-targeting

If one of the biggest companies on the planet—with help from world-class media-buying agencies—couldn't throw money at social media and succeed, then no one can.

features, or the new kid on the block, TikTok. Inside each of these platforms are targeting features that enable zeroing in on your audience with a degree of precision hard to fathom in the presocial era. Want to reach soccer moms in California who drive Teslas? Facebook can help.

Then, ask what result you want to achieve. Are you trying to build awareness for a new product, attract traffic to your website, compile leads, or actually drive conversions? Each of these goals requires a different approach, with different kinds of content and ad tools. The all-familiar sales funnel that has informed marketing and sales for generations—the classic push from awareness to consideration to conversion—holds true in the social space, and its nuances need to be respected.

Finally, ensure you understand how to interpret the results from social ad spend. CPI (cost per impression), CPC (cost per click), and CPE (cost per engagement) might seem like an alphabet soup at first glance, but knowing these terms is critical. Are you spending ten dollars just to get a consumer to click on a social post selling a five-dollar widget? That math simply doesn't work out. Advertising pioneer John Wanamaker famously said way back in the early twentieth century that "half the money I spend on advertising is wasted. The trouble is, I don't know which half." The beauty of social media is you can see exactly which ad spend is driving which result—if you know where to look and what to look for.

Here's where things get a little more complicated, though ... and a little more interesting. In today's landscape, to elevate social from a tactical channel to a growth engine requires more than just traditional paid ads. You need to call in the influencers.

For the uninitiated, an "influencer" is a subject area expert on social media who has accrued a sizable following and level of legitimacy. For every conceivable topic—from beauty tips and Barbie dolls to software hacks, and race cars—there are influencers out there. They come in different forms (kidfluencers, gaming influencers, or synthetic/virtual influencers) and sizes (micro or nano, celebrities in their own right).

Influencers work for a very simple (and powerful) reason: people trust people. A study by Forrester found that the main driver of awareness for customers in social media was the activity of their social connections.[16] In fact, they found that 81 percent of customers who credited social media for making them aware of a recent purchase reported it was either the passive action of seeing a friend's post (52 percent) or actively asking their connections (27 percent) about a product or service in which they were interested.

And if you're skeptical about the power of the influencer, listen to the tale of the Super Puff.

If you're not familiar with the Super Puff, it's a toasty, oversize winter jacket sold by a company called Aritzia (which, in the spirit of full disclosure, I sit on the board

16 "Why Search + Social = Success for Brands," Forrester Consulting, April 2016, https://www.catalystdigital.com/wp-content/uploads/WhySearchPlusSocialEqualsSuccess-Catalyst.pdf.

of). When Aritzia launched its own version of the puff jacket, they turned to paid influencers to spread the word. And not just any old influencers. Aritzia hitched its wagon to the Kardashian brand and spent a chunk of its ad budget on Instagram posts featuring Kendall Jenner.

Now you don't associate Kendall Jenner with winter gear—she's a California girl, after all—but you *do* associate her with over one hundred million Instagram followers and setting trends in mainstream fashion, so it was little surprise that the plan worked. Brian Hill, Aritzia's founder, noted, "Within a day, we acquired 10,000 new Instagram followers and our rate of sale for this item increased by over 700 percent."

Please, for a moment, consider those numbers.

Ten thousand new Instagram followers.

Seven hundred percent sales growth.

In less than twenty-four hours.

Those are numbers that P&G can only dream of, numbers so mind-blowing that Aritzia enlisted Meghan Markle and Bella Hadid for the company's ensuing influencer campaigns. The concept behind influencers is nothing new—it's simply the idea of third-party cred-ibility. An old-school example would be what Michael

Jordan did for Nike. Before launching the Air Jordan brand in 1985, Nike was known as a well-performing running gear company, and that's pretty much it. Post-Air, it became a cultural touchstone.

Thanks to social media, this tactic is no longer limited to iconic brands and iconoclastic celebrities. Today, companies of all sizes can pay influencers of all sizes to "vouch for" their quality, credibility, and cachet. The underlying principle remains the same, however: the covert power of influence and magnetism rather than the brute force of the hard sell. On her Aritzia IG posts, Kendall wasn't outright saying, "This jacket is awesome; buy it;" nor was MJ saying in his Nike ads, "My shoes are awesome; buy them." Jenner simply wore the jacket in some posts, and Jordan just jumped really high in an iconic series of television commercials and print ads. But by appearing with the product, they were saying, "I like this stuff." And if they like the stuff, many consumers will like the stuff … or, at the very least, these ads will put the companies on a consumer's personal radar.

Nor do all influencers have to be of the Kardashian variety, household names with millions of followers. The right microinfluencer can be just as effective, even if they count just a few thousand highly focused followers. For instance, Swedish watchmaker Daniel Wellington relies almost exclusively on microinfluencer campaigns to drive sales. The company does exactly zero traditional advertising, choosing to spend its budget on multiple influencers. Thanks to this philosophy, as of this writing, its visibility is at the highest point in the company's history.[17]

I wish I could lay out a blueprint for a successful influencer or microinfluencer campaign, but tactics are fluid and evolving fast. Ultimately, success comes down to empowering your brand team—the people who know your customers best—with the social ads budget and creative freedom to cut through the noise of social media. Brands that have the most success with influencer campaigns defy convention and predictability. They'll use multiple pitch people, boost their posts at seemingly illogical days and times, and get into metaphorical bed with a personality who seems to have little or nothing to do with their brand or product. For example, asking DJ Khaled to hype chewing gum doesn't seem the least bit

---

17   Lauren Moreno, "10 Great Examples of How Brands Are Leveraging Micro-influencers," Social Media Strategies Summit blog, August 1, 2019, https://blog.socialmediastrategiessummit.com/10-great-examples-of-how-brands-are-leveraging-micro-influencers/.

logical, but when Stride gum hired the hip-hop mogul for a Snapchat takeover, it sold forty-three thousand items that very day.[18]

Influencer campaigns will perhaps become even more pertinent as social networks and e-commerce functionality become more innately linked. The influencer marketing industry is already on track to be worth $15 billion by 2022, up from $8 billion in 2019, according to Business Insider's 2020 *Influencer Marketing Report*. It's safe to say influencer and microinfluencer models aren't going anywhere for a good, long while, so you'll probably want to join the party. Now.

---

18   Christine Smith, "Successful Influencer Marketing Examples in 2019," Influenex, October 17, 2019, https://www.influenex.com/blog/successful-influencer-marketing-examples.html.

PRO TIPS

- Be strategic. Ad hoc advertising postpromotion won't deliver success. To get real results, you need a planned, structured ad strategy that intelligently connects to the rest of your social strategy, advocacy efforts, and other channels such as email and paid media.

- Empower your social teams with a paid advertising budget. Combining organic social media with paid not only can increase your reach, but it will allow you to target more precisely and at scale while optimizing your spending.

- Develop attribution models to quantify results. Attribution still has a long way to go in social media. But borrowing from established models used by paid media teams helps quantify how your social efforts translate to bottom-line impacts.

- Bigger isn't always better. When it comes to influencer marketing, think relevance over reach. A Kendall Jenner–sized follower count is meaningless if her followers aren't interested in your offer. Engagement rates for micro- or nanoinfluencers are often higher than their influencer big brothers and sisters.

CHAPTER 6

The Power of Social Employees

Film has the Oscars. Television has the Emmys. Music has the Grammys.

Social media has the Shortys.

Launched in 2008 by Greg Galant of the New York City–based tech company Sawhorse Media, the Shorty Awards honor "the best of social media by recognizing the influencers, brands, and organizations on Facebook, Twitter, YouTube, TikTok and more." There were dozens of Shortys handed out at the 2020 ceremony, and among the winners were Adidas Runtast-ict (Best Brand Redesign), Netflix and hi5.agency (Best Art Direction), Zendaya (Best Celebrity), and, boooooring, the hand over mouth (Emoji of the Year).

Considering that they came up with an Emoji of the Year award, you're probably wondering how the Shortys came to be. So here's the backstory.

Greg was an early Twitter fan. What he loved, back in those early days, was the camaraderie. "There was a real community on Twitter in a way you don't experience now," he told me in an interview for this book. "No one understood how to use it. You had something in common just by being on it. It felt a lot safer. You could put a half-baked idea out there and know that it would be collaborative." There was just one problem. As Twitter grew, it was difficult to figure out who to pay attention to and whose ideas were worth following.

True to the ethos of the platform, Greg felt that crowdsourcing would be as good a way as any to help decide who and what merited a Twitter user's time. So he decided to create an awards show that could help an overwhelmed user discern the useful stuff from the noise. Over the span of two weekends, Greg slapped together a website for the Shortys, and off he went.

To determine winners, Greg simply asked Twitter users to cast a vote for their favorites, with tweets instead of ballots. The voting concept was an instant hit. Greg figured he'd end up with thirty or so participants, but within twenty-four hours "Shorty Awards" began trending, and the voters numbered in the thousands.

Meanwhile, north of the border in Canada, my little social media management start-up saw what was going on with the Shortys, and we knew we had to win one. Mind you, this wasn't a matter of ego; we just knew the crowdsourced award concept was kind of revolutionary, and nabbing one would lead to some good PR for the company.

The next year, we set out to gamify a victory for our platform. We knew that a lot of our customers were influential on social, so we encouraged them to vote on Twitter and spread the word even further.

Not only that, but we also put our staff on the case. After all, our team was comprised of hardcore social pros, and part of our company culture had become using employees to amplify positive Hootsuite news. We sent out prepopulated scripts so people could share the vote link on their feeds, which led to literally thousands of votes a day.

It worked. We won. By a landslide.

The event itself was impressive, especially considering Greg's modest start. MC Hammer performed (Greg lined him up by sending him a DM on Twitter, which was the definition of apropos), the cosponsors were the Knight Foundation and Pepsi, and among the attendees were New York Times digital blogger Brian Stelter and marketing guru Gary Vaynerchuck.

It was, as the kids like to say, lit.

For both Greg and myself, this was an aha moment: Greg realized that he wasn't the only social media truther out there, and I realized that digital-centric companies needed a more efficient way to harness the power of their employees as brand advocates.

Employee advocacy is an incredibly effective tool to amplify your message. Chances are, your company already has dozens—if not hundreds or even thousands—of champions on the payroll who would love to tell your story and sing your praises to their social followers. Once you start doing the math, you'll understand the untapped resource this represents. A company with one hundred employees, each of whom has an average of five hundred followers, can suddenly reach fifty thousand people. LinkedIn reports that, on average, an organization's employees have ten times more combined followers than the organization itself.[19] Not to mention, Facebook prioritizes posts from friends and family over public content, meaning employee-shared content will spread farther and wider.

Shortly after our Shorty experience, we decided to launch an employee advocacy tool for brands called Amplify, which today remains one of our best-selling products for companies who have chosen to embrace social holistically. Across industries and in companies of all sizes, leveraging the social networks of their employees has proved one of the most effective and cost-efficient ways to get the word out about campaigns, product launches, and company news. Throughout the COVID-19 crisis, for instance, Hootsuite's "Stay Connected" content reached over 268,000 individuals in less than two weeks through employee sharing alone. External advocates might be able to grow your brand, but it's your internal champions that'll make it transformative.

You need quantification on that? Well, according to

---

19   Katie Levinson, "What Is Employee Advocacy and How Do Marketers Win with It?," LinkedIn, March 13, 2018, https://business. linkedin.com/marketing-solutions/blog/linkedin-elevate/2017/ what-is-employee-advocacy--what-is-it-for--why-does-it-matter-.

Edleman's 2019 Trust Barometer report, 53 percent of all global consumers see employees as the most credible source of information about a company, more credible than journalists or industry analysts. On one hand, that wouldn't appear to make sense—you'd expect a third-party entity to be more objective about a company, thus more trustworthy. But on the other hand, if an employee is raving about their workplace or the product their company sells, that says something. Because we've all had workplaces that *definitely* weren't worth raving about.

In short: Your employees can be beacons of trust for your company, and having them share content on social media is a powerful way to get that message out. Deeper still, your employees are likely to be connected to peers who care about the same values, industries, and issues—their affinities map and overlap in rich and complex ways. So your message is not only getting out, it's reaching a very strategic audience.

So how do you evolve your company into exactly this kind of a social-first organization? What's critical to appreciate is that this isn't just a matter of hiring a social media manager for your marketing team. A true social organization has social media integrated throughout the company, in every department, and through every phase of the buyer's journey.

This is paramount because all too often, social media is a fragmented, frustrating experience for consumers, what with marketing, sales, and customer service in separate silos. But when you have a true social org, you can seamlessly guide and support the customer throughout the entire cycle of your relationship. That means that social strategies and tools are part of not just your marketing efforts but also your sales process and customer support initiatives.

> A true social organization has social media integrated throughout the company, in every department, and through every phase of the buyer's journey.

Achieving this kind of social integration means making social media part of your company's DNA. It's not just a tool you use with clients. It has to be part of how you communicate internally and share information as well. Platforms like Slack and Facebook Workplace are showing the power of bringing a social media mindset "inside the firewall." If you can change how your workforce interacts and organizes among themselves, you can change how you communicate with your customer.

Done right, a unified social experience supports customers all the way from building initial awareness for a product to converting that awareness into a purchase, providing customer support, and turning customers into real brand advocates. In a perfect world, the marketing, sales, and customer service teams all have insights and touchpoints with the customer via social media. Information is shared rather than siloed and follows the customer wherever she goes.

The result is that the customer only experiences one unified company—and isn't expected to hunt down the right person or department for different inquiries. This all goes against the way things have been done in business for many, many decades, and it requires a shift in philosophy and approach by businesses themselves. So why embrace all of this now? Why not keep your company in its comfort zone? Why be so disruptive?

Because oftentimes, disruption leads to results.

Take the impact of incorporating social selling into your workflow—i.e., using social media tools to reach out and cultivate prospects. In 2019, the consulting firm Forrester reported that companies who embraced social selling were 72 percent more likely to exceed quotas than their peers who didn't.[20] They added that only 20 percent of companies were equipping their sales teams with social engagement tools, while even fewer figured out how to optimize the practice. Sounds like an opportunity to me.

Social media use can also lead to profound improvements in internal communication and the kind of powerful employee advocacy that goes along with it. We're seeing this during the COVID-19 pandemic in real time, especially with the shift to remote work. More so than ever before, messaging apps like Slack and Microsoft 365, as well as meeting apps such as Zoom and Hangout, are helping leaders maintain useful connections and sustain productivity. Indeed, a recent survey by Deloitte shows that despite all the disruptions caused by the crisis, productivity remained relatively constant, largely thanks to effective use of technology.[21]

---

20   Mary Shea, Jacob Milender, et al., "Add Social Selling to Your B2B Marketing Repertoire," Forrester Consulting, February 7, 2017, https://www.forrester.com/report/Add+Social+Selling+To+Your+B 2B+Marketing+Repertoire/-/E-RES136248#.

21   "How Covid-19 Contributes to a Long-Term Boost in Remote Working," Deloitte, accessed August 2020, https://www2.deloitte. com/ch/en/pages/human-capital/articles/how-covid-19-contributes-to-a-long-term-boost-in-remote-working.html.

But the power of a social org goes beyond just KPIs like productivity. A feeling of belonging is a fundamental human desire, and our work community is a major source of belonging. This can be as simple as a leader asking a member of their team a sincere question about their family or their weekend. That connection builds solidarity, fulfillment and loyalty, and it can be preserved and even strengthened by tapping into social tools.

We've probably all seen examples of this already. There are "water cooler" channels in Slack where employees are encouraged to simply shoot the breeze. How about a virtual team lunch break? Or an online happy hour? Or a daily fifteen-minute Zoom meeting during which no topics are off-limits … except for work?

The common bond between engaged employees, happy customers, and productive companies in 2020 is increasingly social media. Companies that integrate social tools to their fullest, both internally and externally, are able to provide a seamless customer experience while generating genuine buy-in from employees. All of this engenders a virtuous cycle—delighted customers and loyal employees learn, absorb, and communicate on social media, becoming brand champions and employee advocates in their own right. Setting this cycle in motion is far from easy, but companies that invest now are building a brand that will stand the test of time, whatever disruptions lie ahead.

PRO TIPS

- Leverage your network. Not all social advocates have to be bought. Sometimes the best third-party credibility you can gain is from employees and customers. Hootsuite's study of over nine thousand B2B and B2C marketing leaders found that just 30 percent of organizations run customer advocacy initiatives and less than one-quarter use employee advocacy or social selling. Don't miss out on these untapped resources!

- Crawl, walk, run. When it comes to integrating social across your org, start with the areas of the business you know your customers, employees, and shareholders will benefit most from, like customer service or sales. As you start to build up expertise internally, it will become easier to adapt social into other areas of the business and increase your social maturity as a company.

- Secure buy-in from the top. Being a social-first organization requires having social leadership. Executives have to buy in to the value of social, and not just externally but internally. At Hootsuite, my executive team and I offer weekly updates over Facebook Workplace (a secure, internal network), which also allows employees to share sentiment, comment, tag their colleagues, and ask questions.

SOCIAL
FUTURE

CHAPTER 7

The Future of Work

I want to change gears for a moment and take you on a brief trip to the future. Not the distant future, just a mere ten years away—but in that span the way we work and use social tools will have changed markedly. So welcome to the year 2030. Let's talk business.

We're ten years removed from the most tumultuous year in our generation, a year that, considering the pandemic, the polarized political landscape, and the racial justice movement in the United States, changed everything.

When we recall our 2020 business life, we remember it as the year traditional office culture died. Sure, we'd been headed in a remote direction before the virus, but thanks to the shelter-at-home orders that were given on and off for months after the initial outbreak, hybrid and permanent work-from-home cultures emerged at accelerated speed.

During the Shift—as it became known—business leaders had to figure out how to connect with their team in a virtual environment. We lost the opportunity to stop by somebody's desk, look them in the eye, and ask them about their family; to spontaneously share a meal; to grab a coffee and talk about life face-to-face. Work-wise, managers couldn't look over employees' shoulders anymore to see if they were working or online bargain hunting or messing with fantasy football lineups.

But for most people, working from home was perfectly fine, even preferable. Suits, watercooler chitchat, and long commutes were replaced with athleisure, virtual events and the occasional "onsite."

Looking back, we now know the office-to-home transition was kind of inevitable. Unlike in the factory era, we no longer needed to be in one place to have access to complex equipment or machinery. Even before COVID-19, in many offices virtual connections—via

Slack, Google Drive, and chat—were far more critical than physical ones.

Not to say there weren't hurdles along the way, one of which was gauging an employee's success rate. Sure, measuring a salesperson's output was still basically the same: they have a quota, and they either hit it or they don't. But when it came to workers whose output isn't readily quantified—i.e., developers, content creators, HR managers—well, that's when things got difficult.

Enter artificial intelligence performance tracking.

Initially, there was pushback from both supervisors and the supervised. The decision makers didn't want to come across as Big Brother, while the rank and file didn't like the feeling they were being watched. Another issue was that it took a lot of trial and error before all of us were comfortable with the AI itself.

For example, figuring out how to measure a coder's performance wasn't black and white. Are they hitting the keyboard? Check. Are they producing lines of code? Check. Is it quality code? Well … what counts as quality?

Until we nailed down the system, a coder who wrote five thousand lines of lousy code a day was rewarded as much, if not *more*, than a coder who accomplished the same thing in one hundred lines. Eventually, however, the AI caught up. Neural networks figured out *for themselves* which code was most efficient and effective, as algorithms grew smarter and smarter.

And eventually, employees appreciated that this type of measurement system worked far better than the arbitrary performance review mechanisms of the past. Solid workers came to realize they had more chances to grow and succeed if their work was measured tangibly rather than anecdotally. They ultimately wanted to work for an organization that had a good grasp on what their contribution was and then rewarded them for it. Within a few years, it got to the point where most of us forgot what a work environment looked like without AI playing this crucial role.

But all this technology didn't come cheap. Quality AI was and is expensive. Acquiring decent AI became an issue for new companies with small budgets.

As a result, by 2030 young start-ups are largely unable to leverage AI, which sets off a vicious cycle: it's tough to scale without AI, but it's tough to afford AI without having scaled. Dominant companies get more dominant. Upstarts with new, and sometimes better, products and services struggle to compete. (If this sounds familiar, it's because it's a challenge we're already wrestling with as giants like Amazon move light-years ahead of competitors in terms of tech and resources.) In 2030, we're still years away from a solution that will level the playing field.

For most of us, Monday-through-Friday, nine-to-five work schedules have become obsolete. When globally distributed workforces became mainstream, there was less need for employees to keep the same hours in different time zones. With more parents choosing to educate their children in

For most of us, Monday-through-Friday, nine-to-five work schedules have become obsolete.

small homeschool-esque settings, rigid work hours became damaging to employer brands. Instead of working on a fixed schedule, knowledge workers now set their own timetables. Indeed, pay is based less on inputs (hours worked) and more on outputs (deliverables and goals achieved).

Not only do employers allow for flexibility with schedules, but progressive brands now play a key role in childcare as well. During the pandemic, top employers came to realize workers who were taking care of kids all day while also trying to do their jobs were understandably exhausted, unfocused, and unproductive.

At first, many companies were resistant to helping out with childcare, in part because it was a logistical nightmare. *Are we supposed to pay for a nanny? Do we set up our own day care centers and/or schools? Should we have multiple plans for families with multiple kids? How will this impact our bottom line?*

Enter the childcare department, a brand-new addition to the corporate landscape. A central team of childcare experts supported by a network of trained sitters, tutors, and counselors, the childcare department has become by 2030 fixture at start-ups and enterprises alike. While neither cheap to support nor easy to coordinate, this service has become a key factor in attracting and retaining top talent.

Traditional recruiting also went out the window during the Shift, and for many of us looking to build a quality team, that was a relief. The 2020-era interview process was broken. Casual interviews were an arbitrary way to determine whether or not somebody was a good fit for a company, both from a cultural and skills perspective.

LinkedIn was among the first to disrupt the space when it invested heavily in AI for recruitment purposes. Its intelligence could acutely pinpoint not only who had the right aptitude, skill, and value set for a company but also which individuals would collaborate best to create high performance teams. At first, Google, Microsoft, and other leading tech firms applied AI to their recruitment process to rave results. Then, LinkedIn made the technology accessible to anyone, from start-ups to enterprises, who immediately found value in attracting and retaining the right talent.

Automating and streamlining the recruitment process is critical now because most companies are location agnostic: the candidate pool is global and thus hundreds, if not thousands, of times bigger than before the Shift.

And now that we've cracked the language barrier, the pool is growing.

It started with real-time captions in Google Hangouts (2020), a feature that's already transcribing speech in real time today. Instant translation of these captions into hundreds of languages comes online next (2022). Then voice tech catches up, powering seamless, real-time audio translation—i.e., the *Star Trek* universal translator for real (2025). By 2030, you can have a virtual meeting with somebody from Japan or Germany or Mexico or wherever, with streaming audio translations that are 99 percent accurate.

Technology has likewise accelerated the learning curve for new recruits. It previously took weeks, if not months or years, to bring a new employee up to speed. Now thanks to the immersive power of VR experiences, recruits are able to virtually learn their jobs before they start and are good to go within days.

With faster onboarding comes higher expectations. Now that we know employees are ready to roll almost immediately, we look for them to contribute faster than we expected a decade ago. But productivity and output bring up the now controversial topic of compensation.

Should employees still be paid based on hours worked? Or are there more effective ways to reward performance? This has become an issue because efficiency gains mean the forty-hour workweek is no longer standard, and many people are down to a twenty-five-hour week … and even struggle to fill those hours. Some employers insist on cutting pay proportionately; others recognize that output is far more important than hours and devise creative ways to compensate workers based on fulfilling goals and KPIs.

Here are some other notable changes you'll find in 2030:

- Company cultures have become much more diverse and inclusive, with technology better enabling all abilities, geographies, and cultures to understand and connect with one another. Companies and employees alike have become more acclimated to "radical diversity," and a culture of respect is expected by mainstream employers.

- Despite the fact that we've had an effective vaccine for years, we now know pandemics happen, so contact tracing, AI-powered temperature monitoring, and health passports remain commonplace.

- The employer-employee relationship has become less personal and has evolved into one of trust and productivity. There aren't any drinks at the end of the day, but that's OK because in 2020, the system was broken and skewed far too heavily toward interpersonal dynamics. In today's business world, productivity is more of a priority for *everybody* than being buddies.

- This isn't policy or anything, but at most companies, there's considerably less political chatter among coworkers. The world is still polarized, and people have realized getting their work done and heading off to spend time with their family and friends is vastly more important than trying to change a coworker's mind. It's still TBD as to whether this is or isn't a good thing.

I hope you enjoyed that little trip to the future! Twenty thirty is still a decade off, and lots can change in that time, but it's clear to me that the workplace shifts accelerated by COVID-19 are anything but temporary. Indeed, the pandemic may well be remembered as a great leveling-up moment for the world of work, a time when we dropped our old industrial-era habits and fully embraced the potential of new technology. One thing is for sure: the companies that simply seek to "go back to normal" after the crisis won't be around for long. Rather, it's those who use this moment to set a new digital baseline and reinvent their policies and practices who will truly thrive.

CHAPTER 8

Evolution of Platforms into Marketplaces

N ow we return to the present time, a year of radical change on every level. Our interpersonal relationships have changed. Our home lives have changed. Our priorities have changed.

And while it may seem somewhat insignificant in the grand scheme of things, how we shop has changed dramatically as well. During the COVID-19 crisis, countless small businesses turned to social media to connect with their customers in new and creative ways. Indeed, Facebook, stepping up to the plate at a critical time, unveiled a brand-new feature called Shops to allow business owners to quickly bring their wares online.

Suddenly, independent retailers—who may never have had an e-commerce footprint before—had a seamless way to showcase products online, market to Facebook's enormous pool of target customers, process transactions, and even handle logistics and delivery. During the crisis, these businesses have all discovered new audiences, new revenue streams, and new ways of selling their products—all using social media. Meanwhile, their customers have also seen the upsides of social shopping in terms of convenience and accessibility.

It doesn't take a Harvard MBA to realize that after the crisis is over, they're unlikely to simply go back to the old way of doing things. An entire new demographic of business owners and customers has been initiated into the world of social commerce.

Social media companies have been getting into bed with retail for several years, and the relationship is getting steamier by the month.

Social media companies have been getting into bed with retail for several years, and the relationship is getting steamier by the month. Sure, pretty much every retailer under the sun now advertises on social media.

But the connections go far deeper—and get far more complex—than that. Some networks have introduced features to allow you to browse a retailer's catalogue and consummate purchases *without ever leaving your social feed*. Others have taken that a big step further and become retail platforms themselves.

Facebook swims in the retail ocean, while Instagram has made some ripples of its own. Facebook Marketplace has become a Craigslist killer, allowing anyone to easily post goods for sale. It now boasts nearly a billion users. The newly unveiled Facebook Shops takes this a step further, enabling businesses to set up an entire online store on Facebook and Instagram in just a few clicks. They're already making significant inroads into the retail world. And Amazon is probably scared as hell.

While the pandemic has done wonders for Amazon's bottom line, a chunk of that revenue could be usurped down the line as Facebook and Instagram continue to amp up their retail footprint. Indeed, it's not difficult to foresee Instagram evolving into 80 percent social network, 20 percent retail hub. Businesses are already using the platform to advertise. Selling goods in the same place makes sense, especially when you consider that Instagram alone reaches an audience of more than one billion people.

This all got me to wondering if Amazon will ever create its own social network. Considering that Amazon is now heavily dependent on Facebook—i.e., an emerging retail *competitor*—to run its ads, this is hardly far-fetched. At the beginning, the "Amazon network" could be the mirror image of Instagram: 80 percent retail, 20 percent social. Maybe it could be rooted in product reviews. Maybe it could be rooted in purchasing history or something entirely different.

In fact, since as early as 2017, Amazon has had its own influencer program, allowing approved users with sufficiently large social media followings to own an Amazon page where they can recommend specific products. The idea of users being able to connect with one another through these pages or socialize in some other way through Amazon isn't completely out of sight.

Whatever the case, Amazon's goal would be to build a bought-in user base of its own, wean itself off of its dependency on Facebook for ads, and become a one-stop destination for sharing, entertainment, socializing, and buying. Considering they've already got Amazon Prime video to draw on for addictive content, this is not as far-fetched as it may sound.

But here's the thing: if all of these entities start dining on a social/retail combo platter, it might confuse and annoy consumers. Instagram users could get fed up with the increasing number of ads on their feed, while Amazon users would wonder why the retail giant is gumming up their site with adorable cat photos. For ideas on avoiding these problems, we can look to China. As a growing number of platforms in Asia demonstrate, diversifying revenue streams—finding that workable hybrid of advertising and subscription- and transaction-based revenue—may be the surest path to long-term viability and financial success.

As of this writing, Facebook generates upward of 98 percent of its revenue from ads, while for Twitter, the share is around 85 percent.[22] Compare that to Tencent, the Chinese internet services giant that counts more than a billion users across its WeChat and QQ messaging platforms (and is also the first Chinese company to be valued at more than $500 billion). In 2019, Tencent was

---

22   David Ingram, "Facebook Nears Ad-Only Business Model as Game Revenue Falls," Reuters, May 4, 2017, http://reut.rs/2qx9fgy.

earning only 17 percent of its revenue from ads; the rest came from a highly diversified revenue stream including gaming (37 percent), online payments and other fees (23 percent), and value-added services like video subscriptions (24 percent).[23] One result is that while users on Facebook now scroll through countless ads, WeChat Moments (the equivalent of the Facebook News Feed) shows users just two ads per day. It's not hard to see how that might lead to an overall better user experience and greater user loyalty. And in fact, as of July 2020, Twitter is in the early stages of creating a subscription feature for users too.

Other Chinese platforms are getting even more creative. QQ Reading, for instance, allows users to read up to two-thirds of an e-book for free. Readers only need to pay if they're genuinely hooked and want to finish the book. Other authors offer their books for free but include a "tipping option" at the end of each chapter for tips from fifteen cents and up. (That sounds pretty appealing to a burgeoning author such as myself.)

---

23   Connie Chan, "Outgrowing Advertising: Multimodal Business Models as a Product Strategy," Andreessen Horowitz, December 7, 2018, https://a16z.com/2018/12/07/when-advertising-isnt-enough-multimodal-business-models-product-strategy/.

On video platform iQIYI, which has more than five hundred million users, roughly 35 percent of revenue comes from ads.[24] But in contrast to YouTube's ads, those on iQIYI are AI powered and directly related to the video content being viewed—i.e., a makeup tutorial might include ads for lipstick rather than, say, Grammarly. Meanwhile, 40 percent of revenue is derived from iQIYI's eighty million paying members, who subscribe for ad-free viewing and higher-quality video, as well as VIP perks like customizable app skins.[25] The on-screen experience, while admittedly busier than Americans are used to, engages viewers with everything from live comments to GIF creation and even the ability to shop for relevant products while you watch.

Streaming music, too, has been taken to the next level. Tencent Music allows artists to *block* streaming of their new releases. Users have to pay for exclusive access, which generates new revenue for both the performer and the

24 Jeff Loucks, Mark Casey, and Craig Wigginton, "Ad-Supported Video: Will the United States Follow Asia's Lead?," Deloitte, December 9, 2019, https://www2.deloitte.com/us/en/insights/industry/technology/technology-media-and-telecom-predictions/2020/ad-supported-video.html.

25 Ibid.

platform. The app also allows users to buy concert tickets and livestream concerts, creating a kind of 360-degree experience missing on American services like Spotify.

Would any of this work with more Western platforms like Facebook, Instagram, and LinkedIn? Certainly there are nuances to Chinese social media—from high levels of government oversight to the way Chinese users leap-frogged straight to a mobile experience—that need to be accounted for. But my point isn't that these strategies can be imported wholesale and applied by Facebook, Twitter, etc. Rather, what's salient here is the insight that monetization is not a binary matter of ads (Facebook, Twitter, etc.) or subscriptions (Spotify, Netflix, etc.). There are countless permutations—incorporating everything from payments to games and one-off transactions—that better account for consumer tastes and help to capture more lifetime value.

In some ways, we're already seeing this diversification happen in the American context … or, at least, *trying* to happen.

YouTube has been a pioneer in this respect. Subscribers have been able to pay for access to ad-free YouTube since 2015. In 2018, YouTube began allowing creators with more than one hundred thousand subscribers to start charging their users a monthly fee for "channel memberships," which include access to badges, unique emoji, livestreams, and other perks. Meanwhile, the Premieres feature lets content creators build a public landing page with a countdown clock to debut new videos and provide a space for fans to chat live.

Once upon a time, Facebook generated more than 15 percent of its revenue from games and other non-ad sources (nearly $1 billion in 2014).[26] But as concerns over the volatility of the ad market have grown—and justifiably so—the network has begun exploring new monetization options, both for itself and its users.

In 2018, they began testing out paid monthly subscriptions for some groups pages, which allow for access to exclusive content. For example, one of the groups in this small pilot, Grown and Flown Parents, created a paid subgroup dedicated to college prep and counseling, available for thirty dollars a month.

All that said, eyeballs remain the primary currency

---

26   Ryan Holmes, "The Secret to the Success of
     China's Social Networks," *Forbes*, January 18, 2019,
     https://www.forbes.com/sites/ryanholmes/2019/01/18/
     the-secret-to-the-success-of-chinas-social-networks/#1dcffef67f41.

among US-based social networks. The priority remains attracting as many users as possible and keeping them on-site as long as possible, so as to better harvest data and, in turn, target and sell ads. In this light, even initiatives like Marketplace or Facebook's Instant Games platform are really less about generating revenue than just about getting users to stick around longer.

But in the era of "peak ad," that strategy may not be viable forever. Awareness of data—and data privacy—has never been higher among consumers, approximately 60 percent of whom no longer trust social media companies.[27] As ads grow ever more sophisticated and ubiquitous, they're also attracting more scrutiny and backlash. (If you've been chased around the internet by retargeted ads, you know exactly how invasive the ad experience has become.) Relying solely on ads is undeniably bad for the user experience. It's only an amount of time before that has an impact on a network's bottom lines.

For Western social networks, Chinese platforms can offer a more nuanced model. Ads remain important,

27   "Trust Barometer Special Report: Brands and Social Media," Edelman, June 18, 2018, https://www.edelman.com/research/trust-barometer-brands-social-media.

but users can pay to opt out. On-site transactions are seamless. Value-added services, from VIP access to customizable skins, are relevant and plentiful. Above all, the lesson that Chinese platforms offer is that people—across the socioeconomic spectrum—are willing and able to *pay* for a better experience. The billion-plus users of Tencent's various platforms are living proof: there's a better, more sustainable way out there to make money off of social media. Ads aren't everything.

It's impossible to predict exactly how these platforms will change, partly because we don't know what kind of societal changes they'll need to react to. However, taking all of the above into account, I feel comfortable in once again imparting this advice.

Start building your business's social media competencies now. Facebook, Instagram, Twitter, and other platforms are not just the advertising channels of the future but the marketplaces of the future. Businesses who bet otherwise may not be in business for long.

CHAPTER 9

Constant Disruption

I f you'd told thirty-year-old me that in less than five years, I'd be heading up a technology company that would give businesses a virtual window into the conversations their target consumers were having, enable them to personalize their communication at scale, and answer customer queries with the click of a button, I wouldn't have believed you.

As a guy who used to stuff double-sided flyers into the windshields of parked cars to promote his small pizza and paintball businesses, I've experienced firsthand the radical transformation social media has had on how we discover, target, and communicate with customers and all stakeholders.

With all its pitfalls and controversy, social media is a marketer's dream and a huge asset for businesses everywhere. And as quickly as the medium has changed our world, it's still evolving. The only truth we can rely on is that social media will continue to have an immeasurable impact on the way we live, communicate, and do business, now and in the future.

Whether you're on firm footing with the technology or taking your first digital footsteps, the most important takeaway I can give you is to start building your social media muscles now. In the big picture, these are still early days for the integration of social media into business. This is precisely when you want to start building the team, leveraging the tools, and creating the blueprint that will allow you to fully embrace your digital potential. And while differing opinions over a network's role in policing hate speech, misinformation, and inappropriate content will continue, we must not forget these conversations are ultimately what will bring the medium into balance.

When I started my journey with Hootsuite nearly twelve years ago, I saw social media as the next big wave in business, and as an entrepreneur, I jumped on, ready to ride it for as long as it would take me. What I realize now is that social media isn't merely a wave: it's a sea change in business and society. It marks a before-and-after point in communication, and there's no going back.

At its heart, social media is about connection. It has the power to bring us together when we're physically apart, to help us discover people and brands who share our values and beliefs, and to open our minds to new ways of thinking. The opposite is also true, which is why the onus is on all of us, but particularly businesses, to utilize social media the way it was intended and to its fullest potential.

Businesses are among the greatest drivers of change, and the standards they set have strong and lasting ripple effects on society. So plan for the long haul, but do so carefully, remembering that when all is said and done, companies cannot thrive without human connection, which is why social media is such a powerful and permanent part of your businesses journey.

ABOUT THE AUTHOR

Ryan Holmes is the founder and CEO of Hootsuite, the world's most widely used social relationship platform with eighteen-million-plus users. He is the author of the Amazon best-selling guide to social media for leaders, *The $4 Billion Tweet*. He writes about entrepreneurship, technology, and the future of work for *Forbes*, *Fast Company*, and *Inc.* and ranks as a global influencer on LinkedIn and Facebook. For more information:

Ryan.ca

Hootsuite.com

CPSIA information can be obtained
at www.ICGtesting.com
Printed in the USA
BVHW092141271120
593939BV00007B/116